THE PERU SPECIES DISCOVERY

What to know about the 27 New Species in Peru's Alto Mayo Region

JAMES D. JOHN

© 2024 JAMES D JOHN

All rights reserved.

No part of this book may be reproduced, stored in a retrieval system, or transmitted in any form or by any means, electronic, mechanical, photocopying, recording, or otherwise, without the prior written permission of the author, except for brief quotations used in critical reviews or articles.

Disclaimer

This book is intended for informational purposes only. The information contained herein is based on the author's research, analysis, and interpretation of available data at the time of publication. While every effort has been made to ensure the accuracy of the content, the author, publisher, and any associated entities make no guarantees about the accuracy, completeness, or suitability of the information for any particular purpose.

Contents

Introduction 5
Chapter 1: The Expedition to Alto Mayo 13
    Planning and Objectives of the 38-Day Survey 13
    Geographic and Ecological Overview of Alto Mayo 15
    Collaboration with Indigenous Communities and Local Experts 17
Chapter 2: Discoveries in Detail 20
    A. Mammals 20
    B. Fish 25
    C. Amphibians 27
    D. Butterflies 29
Chapter 3: Methodology 33
Chapter 4: Broader Ecological Findings 45
Chapter 5: Conservation Implications 59

    A. The Role of New Discoveries 59

    B. Time-Sensitive Goals 63

Chapter 6: Human and Cultural Dimensions 70

    A. Indigenous Perspectives 70

    B. Strengthening Community-Led Conservation 74

Chapter 7: Challenges and Future Directions 83

    A. Challenges Faced During the Expedition 83

    B. Next Steps in Research and Conservation 87

    C. Raising Awareness 92

Conclusion 97

Introduction

Biodiversity is the cornerstone of a healthy and functioning ecosystem, and the discovery of new species is a testament to the richness and complexity of our planet's natural world. In recent years, environmental conservation efforts have intensified as scientists and researchers strive to understand, document, and protect ecosystems under threat. One such effort led to a groundbreaking expedition conducted by Conservation International in the Alto Mayo landscape of northwestern Peru. This 38-day survey unveiled an astonishing 27 animal species previously unknown to science, ranging from mammals and fish to amphibians and butterflies.

Overview of the Expedition and Its Significance

The expedition, carried out by Conservation International, sought to address critical gaps in biodiversity data within the Alto Mayo landscape, an area spanning approximately 1.9 million acres of forest and agricultural land. Despite being densely populated and heavily influenced by human activity, this region remains a biodiversity hotspot with immense ecological importance. Researchers aimed to not only document species but also understand the health and resilience of these ecosystems in the face of mounting environmental pressures.

Over the course of the expedition, scientists cataloged over 2,000 species, 49 of which are classified as endangered or vulnerable on the International Union for Conservation of Nature

(IUCN) Red List. The discovery of 27 entirely new species was both surprising and significant, highlighting the Alto Mayo landscape as an untapped reservoir of biodiversity. Each discovery contributes to our understanding of how species adapt, survive, and interact within their environments.

Background on Conservation International and the Alto Mayo Landscape

Conservation International (CI) is a global non-profit environmental organization founded in 1987. CI works to protect nature for the benefit of biodiversity, ecosystems, and human well-being. Through science, partnerships, and fieldwork, CI has been instrumental in identifying and safeguarding biodiversity hotspots around the world. The organization emphasizes collaborative approaches, often partnering with local communities, Indigenous groups, governments, and researchers.

The Alto Mayo landscape, located in Peru's San Martín region, represents a crucial ecological corridor bridging the Andes mountains and the Amazon rainforest. Known for its varied ecosystems, the region includes cloud forests, swamps, and river systems, each supporting unique flora and fauna. Despite its ecological importance, the Alto Mayo area faces significant environmental challenges, including deforestation, illegal logging, unsustainable agriculture, and encroachment by human settlements.

In response to these threats, Conservation International has been actively involved in conservation initiatives in Alto Mayo. Efforts include the establishment of protected areas, sustainable agriculture programs, and engagement with Indigenous communities, particularly the Awajún people. By integrating scientific research with community-based conservation strategies, CI aims to create lasting solutions for both people and nature.

Importance of Biodiversity Research in Regions with High Human Influence

Regions like Alto Mayo, where human activity intersects with critical ecosystems, present unique challenges and opportunities for biodiversity conservation. On one hand, human influence often leads to habitat fragmentation, pollution, and resource exploitation, threatening the survival of vulnerable species. On the other hand, these regions also offer valuable insights into how species adapt to changing environments and how conservation efforts can be designed to mitigate human impacts.

Biodiversity research plays a vital role in identifying and documenting species before they are lost to extinction. Each species, whether a tiny insect or a large mammal, plays a specific role in maintaining the balance of its ecosystem. Understanding these roles helps scientists design targeted conservation measures that address the needs of specific species and their habitats.

Additionally, biodiversity research in human-influenced regions contributes to global conservation goals, such as those outlined in the United Nations Sustainable Development Goals (SDGs) and the Convention on Biological Diversity (CBD). By providing data on species distribution, population trends, and ecological interactions, research expeditions like the one conducted in Alto Mayo inform evidence-based policymaking and resource management.

Furthermore, these expeditions foster collaboration between scientists, local communities, and Indigenous knowledge holders. The Awajún people, who participated in the Alto Mayo expedition, provided invaluable insights into the region's ecosystems, traditional conservation practices, and cultural significance of certain species. Such collaborations demonstrate that effective conservation is not solely a scientific endeavor but a shared responsibility involving multiple stakeholders.

In conclusion, the Alto Mayo expedition conducted by Conservation International underscores the urgent need for continued biodiversity research, particularly in regions under significant human influence. The discoveries made during this survey not only expand our scientific understanding but also serve as a clarion call for increased conservation efforts. Through collaboration, innovation, and sustained commitment, it is possible to protect these invaluable ecosystems for future generations.

Chapter 1: The Expedition to Alto Mayo

Planning and Objectives of the 38-Day Survey

The 38-day expedition undertaken by Conservation International in the Alto Mayo landscape was a meticulously planned initiative aimed at addressing significant gaps in biodiversity data. The planning phase involved extensive research, logistical coordination, and collaboration with local authorities and Indigenous communities. Scientists and researchers set clear objectives, including identifying new species, assessing the health of ecosystems, and documenting threats posed by human activity.

One of the primary goals of the expedition was to conduct a Rapid Assessment Program (RAP), a methodology designed to quickly gather comprehensive biodiversity data in remote or under-studied regions. This approach allows scientists to document species diversity, ecosystem health, and potential conservation priorities within a limited timeframe. The Alto Mayo landscape was chosen specifically because of its ecological importance and the pressing need for conservation interventions.

The team comprised biologists, ecologists, taxonomists, and local Indigenous experts, each bringing specialized knowledge to the survey. Equipped with state-of-the-art tools, including camera traps, audio recording devices, and sampling equipment, researchers meticulously surveyed different ecosystems across the region.

The team also prioritized knowledge sharing and capacity building, ensuring that local participants gained skills and experience in biodiversity monitoring.

Geographic and Ecological Overview of Alto Mayo

The Alto Mayo landscape is situated in the San Martín region of northwestern Peru and spans approximately 1.9 million acres. It serves as a critical link between the Andes mountain range and the Amazon rainforest, forming a mosaic of ecosystems that support extraordinary biodiversity. The region's elevation gradient creates diverse habitats, including cloud forests, lowland rainforests, wetlands, and river systems.

Cloud forests in Alto Mayo are characterized by persistent mist, which provides moisture essential for a variety of plant and animal species. These forests are home to endemic flora and fauna, including rare orchids, amphibians, and birds. In contrast, the lowland rainforests are teeming with biodiversity, hosting mammals, reptiles, and a vast array of insect species.

Despite its ecological wealth, the Alto Mayo region faces numerous environmental threats. Deforestation driven by illegal logging and agricultural expansion has led to significant habitat loss. Swamp forests, critical habitats for amphibious rodents like the newly discovered semi-aquatic mouse, are particularly vulnerable. Addressing these threats requires both scientific knowledge and community-driven conservation strategies.

Collaboration with Indigenous Communities and Local Experts

A defining aspect of the Alto Mayo expedition was its collaboration with Indigenous communities, particularly the Awajún people, who have lived in the region for generations. Their deep connection to the land, coupled with traditional ecological knowledge, played a crucial role in the success of the survey.

Indigenous researchers worked alongside scientists, sharing insights into animal behavior, plant identification, and ecosystem dynamics. The partnership was mutually beneficial, as it also empowered local participants with scientific tools and knowledge to monitor and protect their natural resources.

The expedition emphasized respectful engagement, ensuring that Indigenous voices were included in every phase of the research. Local experts were trained in data collection, species identification, and biodiversity monitoring techniques. This capacity-building approach not only strengthened the scientific outcomes of the expedition but also fostered long-term stewardship of the region's natural resources.

In summary, the 38-day expedition to Alto Mayo was a monumental achievement in biodiversity research. It combined scientific rigor, Indigenous knowledge, and collaborative efforts to uncover hidden ecological treasures. The findings from this survey serve as a powerful reminder of the importance of conservation in regions where human activity intersects with nature's delicate balance.

Chapter 2: Discoveries in Detail

A. Mammals

1. Amphibious Mouse

One of the most remarkable mammalian discoveries in Peru's Alto Mayo region is a species of amphibious mouse distinguished by its webbed toes, which adapt it for semi-aquatic life. This unique characteristic allows the mouse to navigate swampy environments and shallow water bodies effectively, a rare trait among terrestrial rodents. The mouse's fur is slightly water-repellent, aiding in buoyancy and insulation while swimming. Additionally, its diet likely includes aquatic insects, small crustaceans, and plant material, though further studies are needed to confirm these details.

The amphibious mouse was found in a small patch of swamp forest, a habitat increasingly threatened by agricultural expansion and deforestation. This specific area is vital for its survival, as the species' semi-aquatic lifestyle limits its ability to adapt to drier environments. Conservationists have flagged this habitat as a high priority for protection due to its ecological uniqueness and vulnerability.

The amphibious mouse underscores the importance of preserving wetlands, not only for their inherent biodiversity but also for the ecosystem services they provide, such as water filtration and carbon storage. Protecting this mouse's habitat would also benefit numerous other species that rely on the same ecosystem.

2. Spiny Mouse

Another fascinating discovery was a species of spiny mouse. These rodents are characterized by their coarse, spiny fur, which serves as a deterrent to predators. The spiny mouse found in the Alto Mayo region has unique coloration and fur texture, suggesting adaptations specific to its environment. This species likely plays a critical role in seed dispersal, as spiny mice are known to forage for fruits and nuts, inadvertently aiding in the regeneration of forested areas.

The ecological role of the spiny mouse extends to its interactions with predators and competitors, forming a vital link in the local food web. However, the species' habitat is under pressure from human activities, highlighting the need for further research and conservation measures to ensure its survival.

3. Short-Tailed Fruit Bat

The short-tailed fruit bat discovered in the survey adds to the known diversity of bat species in the region. This small, agile flyer is integral to the ecosystem due to its role in pollination and seed dispersal. Feeding primarily on fruits and nectar, the short-tailed fruit bat helps maintain the health and diversity of the Alto Mayo's forests by facilitating the reproduction of various plant species.

The bat's short tail and distinctive facial markings set it apart from other species in its genus. Despite its ecological importance, bats often face threats from habitat loss and negative public perceptions. Protecting this newly discovered species involves raising awareness of its crucial role in forest ecosystems and implementing measures to safeguard its roosting sites.

4. Dwarf Squirrel

The discovery of a dwarf squirrel species in the Alto Mayo region highlights the adaptability and diversity of mammals in this landscape. This small, elusive creature is likely arboreal, relying on the forest canopy for shelter and food. Its diet probably includes seeds, nuts, and insects, making it a key player in both seed dispersal and pest control.

Dwarf squirrels are often sensitive to habitat changes due to their dependence on intact forest canopies. The ongoing deforestation in the region poses a significant threat to their survival. Conservation efforts aimed at preserving forested areas will be crucial for the dwarf squirrel and many other species that share its habitat.

B. Fish

1. The "Blob-Headed" Fish

One of the most unusual discoveries during the expedition was the "blob-headed" fish, named for its distinctive, enlarged head. This feature, which resembles a swollen blob or oversized nose, is unlike anything previously documented in ichthyology. Scientists theorize that the head structure may enhance the fish's ability to detect food, possibly functioning as a sensory organ. However, the exact purpose of this feature remains a mystery, sparking significant scientific interest.

The blob-headed fish was found in a fast-flowing stream within the Alto Mayo landscape. Its habitat preferences suggest it might be adapted to specific water conditions, such as high oxygen levels and clear waters. Protecting these freshwater ecosystems is essential for preserving the blob-headed fish and other aquatic species that are yet to be discovered.

2. Additional Fish Species

Beyond the blob-headed fish, the expedition uncovered seven other fish species new to science. These include a variety of small, colorful fish adapted to diverse aquatic habitats, from mountain streams to swampy lowlands. Each species exhibits unique morphological traits, such as specialized fins, coloration, or feeding mechanisms, reflecting the ecological richness of the region.

These discoveries highlight the importance of freshwater habitats within the Alto Mayo landscape. Rivers and streams not only support a wide range of fish species but also serve as critical water sources for surrounding ecosystems and human communities. Conservation strategies must address threats such as pollution, overfishing, and habitat modification to safeguard these vital aquatic ecosystems.

C. Amphibians

The survey also identified three new species of amphibians, each with distinctive traits and ecological roles. Amphibians are particularly sensitive to environmental changes, making them important indicators of ecosystem health.

1. Species 1

One of the newly discovered amphibians is a brightly colored tree frog with striking patterns on its skin. These patterns likely serve as a warning to predators about the frog's toxicity. The tree frog was found in a high-altitude cloud forest, where it likely plays a role in controlling insect populations.

2. Species 2

Another amphibian species is a ground-dwelling toad with cryptic coloration, allowing it to blend seamlessly with leaf litter. This adaptation helps it avoid predators and hunt small invertebrates. The toad's habitat is a lowland rainforest area, emphasizing the need for habitat-specific conservation efforts.

3. Species 3

The third amphibian is a small, slender salamander with semi-aquatic habits. It inhabits streams and nearby moist areas, relying on its permeable skin for respiration. This species' presence underscores the importance of clean, unpolluted water sources for amphibian survival.

D. Butterflies

The expedition documented 10 new butterfly species, further illustrating the Alto Mayo's extraordinary biodiversity. Butterflies are not only visually stunning but also play critical roles in pollination and ecosystem health.

1. Morphological Diversity

The newly discovered butterflies exhibit a wide range of wing patterns and colors, which likely serve various ecological functions, such as camouflage, mate attraction, and predator deterrence. Some species display iridescent scales, which may help regulate body temperature or enhance visual signaling.

2. Pollination and Ecosystem Health

Butterflies are essential pollinators, particularly for plants that require specialized pollination mechanisms. By transferring pollen as they feed on nectar, butterflies support the reproduction of numerous flowering plants, contributing to the region's floral diversity. Additionally, they serve as prey for various predators, forming an integral part of the food web.

3. Conservation Importance

The discovery of these butterfly species emphasizes the need to protect their habitats from deforestation and pesticide use. Butterfly populations are highly sensitive to environmental changes, making them valuable indicators of ecosystem health. Conservation initiatives that prioritize the preservation of forests and sustainable agricultural practices will be crucial for maintaining butterfly diversity.

The discoveries of new mammal, fish, amphibian, and butterfly species in the Alto Mayo region underscore the unparalleled biodiversity of this landscape. Each species contributes uniquely to its ecosystem, highlighting the intricate interdependence of life forms. Protecting these species and their habitats is vital for preserving the ecological balance and ensuring the survival of countless other organisms that depend on these ecosystems.

Chapter 3: Methodology

A. The Rapid Assessment Program

The Rapid Assessment Program (RAP) is a pioneering initiative spearheaded by Conservation International to evaluate biodiversity in regions that are often underrepresented in scientific studies. This program, under the leadership of Trond Larsen, has been instrumental in providing critical data on ecosystems across the globe. The 38-day expedition in the Alto Mayo landscape of northwestern Peru exemplifies the RAP's commitment to uncovering unknown species and assessing the health of ecosystems in a time-sensitive manner.

Goals of the Rapid Assessment Program

The overarching aim of the Rapid Assessment Program is twofold:

1. **Identifying Unknown Species:** The RAP focuses on cataloging species that have either never been scientifically documented or whose ecological roles remain poorly understood. These species often inhabit remote or ecologically fragile regions, making them particularly susceptible to habitat loss and environmental changes. By identifying such species, the RAP aims to enrich global biodiversity records and prioritize conservation efforts.
2. **Gathering Conservation Data:** Beyond species identification, the program seeks to assess the overall health of ecosystems. This involves studying species

interactions, population dynamics, and the impact of human activities such as deforestation and agricultural expansion. The data gathered is essential for developing evidence-based conservation strategies tailored to the specific needs of each region.

Fieldwork Approach

The RAP employs a highly systematic approach to fieldwork. Teams of scientists, ecologists, and local collaborators conduct intensive surveys in targeted regions. In the Alto Mayo expedition, the RAP team focused on a diverse landscape spanning from the Andes to the Amazon. This area's unique topography and ecosystems provided a fertile ground for discoveries.

Key steps in the fieldwork included:

- **Preliminary Planning:** Before setting foot in the field, the team conducted thorough research on the Alto Mayo region. This involved studying existing biodiversity data, consulting with local communities, and identifying potential areas of interest.
- **Species Inventory:** Using a combination of traditional techniques and modern tools, the team cataloged flora and fauna. Methods included camera trapping, audio recording for vocal species, and DNA sampling to confirm genetic uniqueness.
- **Environmental Monitoring:** The team collected data on soil quality, water chemistry, and other environmental parameters to understand how local ecosystems function and how they are influenced by external factors.

Results of the Alto Mayo Expedition

The RAP's work in Alto Mayo was groundbreaking. The team documented 27 species new to science, including mammals, fish, amphibians, and butterflies. Among these, the amphibious mouse with webbed toes and the blob-headed fish with an enigmatic cranial structure stood out as remarkable discoveries. These findings underscore the importance of rapid biodiversity assessments in highlighting the ecological richness of underexplored regions.

B. Indigenous Collaboration

The collaboration with indigenous communities, particularly the Awajún people, was a cornerstone of the Alto Mayo expedition. Conservation International recognizes that sustainable conservation efforts must integrate the knowledge, values, and participation of local communities who have lived in these regions for generations.

Role of the Awajún Community

The Awajún people are an indigenous group with a deep connection to the Alto Mayo region. Their traditional knowledge and practices have evolved over centuries, offering unique insights into the local ecosystems. Their role in the expedition included:

- **Guiding Fieldwork:** The Awajún's familiarity with the landscape proved invaluable for navigating remote and challenging terrains. They helped identify areas of ecological significance that might otherwise have been overlooked.

- **Species Identification:** With their intimate understanding of local flora and fauna, the Awajún provided critical information on species behaviors, habitats, and interactions. This knowledge complemented scientific observations and enriched the overall dataset.

Insights from Indigenous Cosmovision

The Awajún cosmovision—a holistic worldview that emphasizes the interconnectedness of all living and non-living elements—played a significant role in shaping the expedition's approach to conservation. Key aspects of their cosmovision include:

1. **Nature as a Living Entity:** The Awajún view nature as a living, sacred entity. This perspective fosters a sense of stewardship

and respect for the environment, which aligns with modern conservation principles.

2. **Sustainable Practices:** Their traditional practices, such as rotational farming and selective hunting, demonstrate a sustainable approach to resource use. These methods can serve as models for balancing human needs with ecological preservation.

3. **Cultural Significance of Biodiversity:** Many species hold cultural or spiritual significance for the Awajún. For instance, certain plants are used in traditional medicine, while specific animals are featured in myths and rituals. Recognizing these cultural dimensions adds depth to conservation efforts.

Examples of Collaboration in Fieldwork

The partnership between the RAP team and the Awajún community extended beyond logistical support to a genuine exchange of knowledge and expertise. Examples of this collaboration include:

- **Joint Species Surveys:** Awajún participants worked alongside scientists to conduct species surveys, blending traditional observation techniques with scientific methodologies. This approach enhanced the accuracy and comprehensiveness of the findings.
- **Conservation Education:** The RAP team conducted workshops to share scientific insights with the Awajún community. These sessions covered topics such as

biodiversity monitoring and the impacts of climate change, fostering mutual learning.

- **Capacity Building:** To empower the Awajún as conservation leaders, the expedition included training programs on data collection, species identification, and habitat restoration. These skills enable the community to continue conservation efforts independently.

Challenges and Successes

Collaborating with indigenous communities is not without challenges. Language barriers, cultural differences, and differing priorities can complicate interactions. However, the Alto Mayo expedition demonstrated that respectful and inclusive engagement can yield significant benefits.

The success of this collaboration is evident in the Awajún's increased involvement in conservation initiatives. For instance, they have begun advocating for the protection of swamp forests critical to the survival of the newly discovered amphibious mouse. Their efforts highlight the potential for indigenous communities to serve as key allies in global conservation efforts.

The Broader Impact of Indigenous Collaboration

By integrating indigenous knowledge and values into scientific research, the Alto Mayo expedition set a precedent for future conservation projects. This approach not only enhances the quality of biodiversity data but also promotes equitable and culturally sensitive conservation practices. Moreover, it empowers indigenous communities to assert their rights and protect their territories, ensuring that conservation benefits both people and nature.

The methodology employed during the Alto Mayo expedition reflects a model of inclusive and impactful conservation. The Rapid Assessment Program's scientific rigor, combined with the Awajún community's traditional knowledge, underscores the importance of interdisciplinary and intercultural collaboration in addressing the global biodiversity crisis. Together, these efforts provide a blueprint for achieving sustainable conservation outcomes in even the most challenging environments.

Chapter 4: Broader Ecological Findings

A. Overall Biodiversity

The recent expedition in the Alto Mayo region of northwestern Peru unveiled a breathtaking level of biodiversity, with a total of 2,000 species documented. Among these, 27 were identified as entirely new to science, demonstrating the ecological richness of the region and underscoring its global importance as a biodiversity hotspot. The findings encompass a wide range of taxa, including mammals, amphibians, fish, and insects, reflecting the interconnectedness and complexity of the region's ecosystems.

The presence of 49 species listed on the International Union for Conservation of Nature's (IUCN) Red List highlights the region's critical role in the survival of vulnerable and endangered species. These include species at various levels of risk, from vulnerable to critically endangered, emphasizing the fragility of these populations. The Alto Mayo region's diverse habitats—spanning from the Andes to the Amazon—provide refuge for species that may not survive elsewhere. This unique ecological continuum supports endemic species, many of which are highly specialized and dependent on particular environmental conditions.

The comprehensive cataloging of biodiversity in this expedition contributes significantly to global conservation science. By providing baseline data, researchers can better understand the distribution, abundance, and ecological roles of species in the Alto Mayo landscape. This knowledge is essential for crafting effective conservation strategies, prioritizing areas for protection, and assessing the impacts of human activities on biodiversity.

Among the standout discoveries, the amphibious mouse and the "blob-headed" fish illustrate the remarkable adaptations of species in this region. Such findings not only enrich our understanding of evolutionary processes but also reinforce the importance of conserving habitats where such unique species thrive. The identification of 10 new butterfly species further underscores the region's ecological importance, as butterflies often serve as indicators of ecosystem health.

The Alto Mayo's biodiversity also holds intrinsic value, providing ecosystem services that benefit both nature and humans. From pollination and seed dispersal to water filtration and carbon sequestration, these services are vital for maintaining ecological balance and supporting human livelihoods. Documenting and understanding the full range of these services will be crucial for ensuring the sustainable use of the region's natural resources.

B. Threats to Biodiversity

Despite its ecological significance, the Alto Mayo region faces severe threats from human activities, particularly deforestation and agricultural expansion. These practices have led to significant habitat loss and fragmentation, disrupting ecosystems and placing many species at risk.

1. Deforestation

Deforestation in the Alto Mayo region is driven by several factors, including illegal logging, land conversion for agriculture, and infrastructure development. The clearing of forests not only destroys habitats but also contributes to climate change through the release of stored carbon dioxide. The region's forests, which act as carbon sinks, play a crucial role in mitigating global warming. Their loss exacerbates climate change impacts, which in turn further threaten biodiversity.

Deforestation also has cascading effects on the region's ecosystems. Many species in the Alto Mayo are highly specialized, relying on specific forest types or microhabitats. The loss of these habitats can lead to population declines and even local extinctions. Additionally, deforestation disrupts ecological interactions, such as pollination and seed dispersal, which are essential for maintaining forest health and regeneration.

2. Agricultural Practices

Agricultural expansion is another major threat to biodiversity in the Alto Mayo region. Traditional farming practices, often involving slash-and-burn techniques, result in soil degradation and loss of biodiversity. The conversion of forested areas into monoculture plantations further reduces habitat complexity, making it difficult for many species to survive.

The impacts of agriculture extend beyond habitat loss. The use of pesticides and fertilizers can contaminate water sources, affecting aquatic ecosystems and the species that depend on them. For example, the newly discovered amphibious mouse, which inhabits swamp forests, is particularly vulnerable to water pollution. Moreover, agricultural practices can introduce invasive species, which compete with native species and disrupt ecosystem balance.

3. Combined Impacts

The combined impacts of deforestation and agricultural expansion are particularly severe in areas where habitats are already fragmented. Fragmentation isolates populations, reducing genetic diversity and making species more susceptible to disease and environmental changes. It also creates edge effects, altering microclimates and exposing forest interiors to higher temperatures and light levels, which can be detrimental to certain species.

C. Conservation Opportunities

The findings from the Alto Mayo expedition provide a unique opportunity to design and implement targeted conservation strategies. By leveraging the data collected, researchers and conservationists can address the region's threats while promoting sustainable development that benefits both nature and people.

1. Protected Areas and Habitat Restoration

One of the most effective strategies for conserving biodiversity is the establishment and expansion of protected areas. The Alto Mayo region already includes several protected zones, but the identification of new species and critical habitats highlights the need for additional protections. Expanding protected areas to include the habitats of newly discovered species, such as the amphibious mouse and blob-headed fish, will be crucial for their survival.

Habitat restoration is another important conservation tool. Reforestation and the restoration of degraded lands can help reconnect fragmented habitats, providing corridors for species to move and adapt to changing conditions. These efforts can also enhance ecosystem services, such as carbon sequestration and water regulation, benefiting local communities.

2. Community-Led Conservation

Engaging local communities in conservation efforts is essential for achieving long-term success. The involvement of indigenous groups, such as the Awajún people, in the Alto Mayo expedition demonstrates the value of traditional ecological knowledge. Indigenous communities often have a deep understanding of local ecosystems and can play a key role in protecting them.

Community-led conservation initiatives can include sustainable resource management, agroforestry practices, and ecotourism. These approaches not only help conserve biodiversity but also provide economic benefits, reducing the pressure to exploit natural resources unsustainably. Supporting indigenous rights and land tenure can further strengthen conservation efforts, ensuring that communities have a stake in preserving their natural heritage.

3. Policy and Research Integration

The data collected during the expedition can inform policy decisions at local, national, and international levels. For example, incorporating the findings into land-use planning can help balance development with conservation. Policies that promote sustainable agriculture, reduce deforestation, and protect critical habitats will be essential for safeguarding the region's biodiversity.

Ongoing research is also vital for addressing conservation challenges. Long-term monitoring of species populations and ecosystem health will provide insights into the effectiveness of conservation measures and allow for adaptive management. Collaborative research initiatives, involving both scientists and local communities, can enhance understanding and foster a shared commitment to conservation goals.

4. Climate Change Mitigation and Adaptation

The Alto Mayo region's forests play a critical role in mitigating climate change, and conserving them is an essential part of global climate strategies. Reforestation and forest conservation efforts can sequester carbon, helping to offset greenhouse gas emissions. Additionally, maintaining healthy ecosystems can enhance resilience to climate change impacts, such as extreme weather events and shifting rainfall patterns.

Adaptation strategies, such as protecting climate refugia and promoting species' adaptive capacity, will also be important. By identifying and safeguarding areas that are likely to remain stable under future climate conditions, conservationists can provide safe havens for vulnerable species. Supporting genetic diversity and ecosystem connectivity will further enhance the ability of species and ecosystems to adapt.

5. Education and Awareness

Raising awareness about the importance of the Alto Mayo region's biodiversity can inspire action at multiple levels. Educational programs, targeting both local communities and the broader public, can highlight the value of conserving the region's natural heritage. Public campaigns and partnerships with media outlets can also help garner support for conservation initiatives and funding.

By sharing the stories of newly discovered species and the efforts to protect them, researchers can foster a sense of wonder and responsibility for the natural world. Engaging the next generation in conservation through school programs and citizen science projects can build a lasting commitment to preserving biodiversity.

The Alto Mayo expedition's findings underscore the urgency of conserving the region's biodiversity in the face of mounting threats. By taking a comprehensive approach that integrates scientific research, community engagement, and policy action, it is possible to protect this unique and vital landscape for future generations. The discoveries made during this survey serve as both a testament to the richness of life on Earth and a call to action to safeguard it.

Chapter 5: Conservation Implications

A. The Role of New Discoveries

Scientific Value of Identifying Unknown Species

The identification of unknown species plays a foundational role in understanding and protecting the Earth's biodiversity. Each newly discovered organism contributes critical data to the global scientific repository, enriching our knowledge of ecosystems, evolutionary processes, and ecological balance. Species, both large and small, are interlinked within intricate webs of life; therefore, understanding one species often provides insights into the broader systems in which it exists.

For instance, the discovery of the amphibious mouse in Peru's Alto Mayo region, with its webbed toes adapted for a semi-aquatic lifestyle, offers a glimpse into the evolutionary pressures and ecological dynamics unique to swamp forests. By studying such adaptations, scientists can infer patterns of environmental change, such as shifts in water availability or vegetation due to climate change or human activity. Similarly, the blob-headed fish, with its unusual morphology, raises questions about sensory evolution and feeding strategies in aquatic environments. Such findings can guide further research into how species adapt to niche habitats under environmental stressors.

New species discoveries are also vital in biotechnology and medicine. Many pharmacological breakthroughs, including life-saving drugs, have roots in natural compounds found in plants, fungi, and animals. Each newly discovered species carries the potential to harbor unique biochemical compounds that could inspire future innovations. Conservationists and researchers emphasize that preserving biodiversity is akin to safeguarding a library of solutions, many of which we have yet to unlock.

Inspiring Global and Local Conservation Efforts

The discovery of new species often generates widespread interest, serving as a rallying point for conservation initiatives. Globally, it underscores the importance of protecting habitats that are home to such biological treasures. For example, the Alto Mayo landscape spans a rich intersection of Andean and

Amazonian ecosystems, and discoveries there highlight the region's unique biodiversity value. By showcasing these findings, conservation organizations can draw attention to the urgent need for preserving these areas.

At the local level, discoveries often foster a sense of pride and stewardship among indigenous and rural communities. In Peru, the collaboration between scientists and the Awajún people exemplifies how involving local stakeholders in research not only enhances scientific outcomes but also strengthens cultural and ecological ties. These communities often possess deep traditional knowledge about their environment, which, when combined with scientific expertise, can create powerful conservation strategies.

Public interest in unique and unusual species also plays a role in driving conservation funding and policy change. Charismatic species like the blob-headed fish or the dwarf squirrel can serve as flagship species for broader conservation campaigns. Highlighting the precarious habitats of such species can galvanize support for habitat preservation, stricter anti-deforestation laws, and sustainable land-use practices.

B. Time-Sensitive Goals

Urgency in Meeting International Biodiversity Conservation Targets

The global biodiversity crisis has reached a critical juncture. Reports from organizations like the Intergovernmental Science-Policy Platform on Biodiversity and Ecosystem Services (IPBES) estimate that over one million species are at risk of extinction due to human activity. Time-sensitive conservation goals, such as the Kunming-Montreal Global Biodiversity Framework, aim to halt biodiversity loss and restore degraded ecosystems by 2030. Achieving these goals requires immediate, coordinated efforts at international, national, and local levels.

In regions like Alto Mayo, where human influence is pronounced, the stakes are particularly high. Agricultural expansion, logging, and infrastructure development have fragmented habitats and pushed many species to the brink of extinction. Rapid assessment surveys, like the one conducted by Conservation International, provide the baseline data needed to identify priority areas for conservation action. The discovery of 49 species on the International Union for Conservation of Nature's (IUCN) Red

List in the region underscores the urgency of implementing protective measures.

Moreover, new species discoveries act as a stark reminder of what is at risk. Species unknown to science are often those most vulnerable to extinction because they typically occupy specialized or isolated habitats. The amphibious mouse, found in a small swamp forest threatened by agricultural practices, is a prime example. Without timely intervention, such species could disappear before their ecological roles are fully understood.

Strategies for Balancing Human Activity and Ecological Preservation

Meeting biodiversity conservation targets requires innovative strategies that balance ecological preservation with human development. One of the most effective approaches is the integration of conservation goals into sustainable land-use practices. In the Alto Mayo region, this could involve promoting agroforestry systems that combine agricultural productivity with habitat conservation. Shade-grown coffee plantations, for instance, can support biodiversity while providing economic benefits to local farmers.

Community-led conservation initiatives are another key strategy. Empowering indigenous communities, like the Awajún, to manage their territories can lead to more effective and culturally appropriate conservation outcomes. These communities have an intrinsic understanding of their environment and a vested interest in its preservation. Providing them with resources, legal recognition, and scientific collaboration can enhance their capacity to protect biodiversity.

Protected areas also play a crucial role in balancing conservation and development. Establishing or expanding reserves in biodiversity hotspots can safeguard critical habitats while allowing for regulated human activity. However, these efforts must be accompanied by measures to address underlying drivers of habitat destruction, such as poverty and lack of access to sustainable livelihoods.

On a broader scale, technological advancements offer new tools for conservation. Satellite imagery and remote sensing can monitor deforestation and habitat loss in real time, enabling quicker responses to threats. Genetic sequencing and environmental DNA (eDNA) analysis allow for the identification of species even in inaccessible areas, accelerating the discovery and protection of unknown biodiversity.

Finally, global cooperation is essential. Conservation is not confined by borders; the impacts of biodiversity loss are felt worldwide. International funding mechanisms, such as the Global Environment Facility, can support conservation efforts in regions like Alto Mayo. Multinational agreements, such as the Convention on Biological Diversity, provide frameworks for collaboration and accountability.

The discoveries in Peru's Alto Mayo region exemplify the dual role of biodiversity research as both a scientific endeavor and a conservation imperative. Identifying unknown species expands our understanding of life on Earth and inspires actions to protect it. However, the window for effective intervention is rapidly closing. By aligning new discoveries with targeted conservation strategies, fostering community involvement, and leveraging global cooperation, we can create a future where biodiversity thrives alongside human development. The stakes are high, but the rewards of success are immeasurable—for both nature and humanity.

Chapter 6: Human and Cultural Dimensions

A. Indigenous Perspectives

Insights from the Awajún People on Biodiversity

The Awajún people, one of the largest Indigenous groups in Peru, inhabit the biodiverse yet fragile landscapes of the Amazon. For centuries, their survival has depended on an intimate understanding of their environment, fostering a deep relationship with the ecosystems they call home. This relationship is built on the premise that all living beings—plants, animals, and humans—are interconnected and hold intrinsic value within the greater web of life.

The Awajún possess extensive traditional ecological knowledge that has been passed down orally through generations. This knowledge encompasses the identification of plants and animals, their uses, and their roles in the ecosystem. For example, the Awajún have identified plants with medicinal properties that are now of interest to scientific research. Similarly, they recognize the importance of certain keystone species in maintaining ecosystem balance, such as seed-dispersing animals that contribute to forest regeneration.

Their insights into biodiversity extend beyond scientific classification. To the Awajún, biodiversity is a manifestation of the sacred order of the universe. Each species is believed to have a spiritual essence, and the loss of any species is viewed not merely as an ecological issue but as a disruption of this sacred order. This perspective underscores the importance of protecting all forms of life as part of their cultural and spiritual heritage.

Cultural Connections to Nature and Ecosystems

The Awajún worldview—or cosmovision—integrates cultural, spiritual, and practical elements of life, rooted in a profound respect for nature. They view themselves not as separate from the environment but as part of it. This connection is reflected in their myths, rituals, and daily practices.

One illustrative example is their agricultural system, which is designed to mimic natural ecosystems. The Awajún practice shifting cultivation, wherein small plots of land are cleared for farming and left to regenerate after a few years. This method allows for sustainable use of resources while preserving forest ecosystems. Similarly, their hunting practices are governed by traditional laws that ensure the continued abundance of game species. Hunters often seek guidance from spiritual leaders before embarking on expeditions, emphasizing the interconnectedness of human activity and natural cycles.

Cultural practices such as festivals and ceremonies further highlight the Awajún's connection to nature. Many of these events are timed to coincide with natural phenomena, such as the flowering of certain plants or the migration of animals. These practices not only celebrate the bounty of the land but also reinforce the community's role as stewards of their environment.

This holistic relationship with nature informs the Awajún's approach to conservation. For them, protecting the forest is not merely about preserving resources for future use but about maintaining the integrity of their culture and spiritual beliefs. This perspective is increasingly recognized as a vital component of effective conservation efforts, as it aligns ecological goals with cultural and social values.

B. Strengthening Community-Led Conservation

Empowering Local Communities Through Scientific Knowledge

Empowering Indigenous communities like the Awajún with scientific knowledge is a critical step toward sustainable conservation. When traditional ecological knowledge is complemented by modern scientific methods, it creates a powerful synergy that enhances understanding and management of biodiversity.

One successful approach is participatory research, where local communities are involved in every stage of scientific investigation—from identifying research priorities to collecting and analyzing data. This method not only validates traditional knowledge but also ensures that scientific findings are relevant to the community's needs. For instance, during the recent survey in the Alto Mayo region, Awajún researchers collaborated with scientists to document species, contributing their expertise in identifying plants and animals. This collaboration led to the discovery of new species and strengthened the community's role as guardians of biodiversity.

Training and capacity-building initiatives are also essential. Workshops and educational programs can provide Indigenous communities with skills in biodiversity monitoring, mapping, and conservation planning. Such programs empower communities to take an active role in managing their natural resources. For example, Awajún youth trained in Geographic Information System (GIS) technology have successfully used satellite imagery to monitor deforestation in their territories, providing critical data for conservation initiatives.

Moreover, integrating scientific knowledge into local education systems can inspire the next generation of conservationists. By teaching children about both traditional and scientific approaches to biodiversity, communities can foster a sense of pride in their heritage while equipping them with tools to address modern environmental challenges.

Examples of Successful Conservation Partnerships

Several initiatives demonstrate the potential of community-led conservation partnerships to protect biodiversity and enhance livelihoods. These examples highlight how collaboration between Indigenous communities, non-governmental organizations (NGOs), and governments can achieve sustainable outcomes.

1. The Alto Mayo Conservation Initiative

The recent biodiversity survey in the Alto Mayo region is a prime example of successful collaboration. Indigenous researchers played a crucial role in the survey, providing insights into local ecosystems and facilitating access to remote areas. The findings, including the discovery of 27 new species, underscore the value of integrating traditional knowledge with scientific expertise. The project's focus on empowering local communities has led to the development of conservation strategies that address both ecological and socio-economic needs.

2. The Awajún Agroforestry Program

To combat deforestation and promote sustainable livelihoods, the Awajún have implemented agroforestry systems that combine tree planting with crop cultivation. Supported by NGOs and government agencies, this program not only restores degraded lands but also provides economic benefits through the production of high-value crops like cacao and coffee. By aligning conservation goals with economic incentives, the program has gained widespread community support.

3. The Indigenous REDD+ Program

The REDD+ (Reducing Emissions from Deforestation and Forest Degradation) initiative has been adapted to include Indigenous perspectives. In the Peruvian Amazon, the Awajún have participated in projects that reward communities for conserving forests, which act as carbon sinks. These projects provide financial incentives while respecting traditional land management practices. The success of the program lies in its recognition of Indigenous land rights and its emphasis on community-led decision-making.

4. The Awajún Women's Conservation Network

Recognizing the vital role of women in conservation, this network empowers Awajún women to lead initiatives such as reforestation and biodiversity monitoring. Through capacity-building workshops and access to microfinance, women have become key players in sustaining their communities' natural resources. This initiative highlights the importance of gender inclusion in conservation efforts.

The human and cultural dimensions of biodiversity conservation are critical to achieving sustainable outcomes. The Awajún people's profound connection to nature and their extensive traditional knowledge provide invaluable insights into preserving biodiversity. By empowering Indigenous communities through scientific knowledge and fostering collaborative partnerships, we can create conservation models that respect cultural values while addressing global environmental challenges.

Ultimately, the integration of Indigenous perspectives and community-led initiatives offers a holistic approach to conservation—one that recognizes the interdependence of people, culture, and ecosystems. This approach not only protects biodiversity but also supports the livelihoods and cultural heritage of the communities who are its stewards. As the

Awajún teach us, the preservation of nature is not just a scientific imperative but a moral and cultural one, essential for the well-being of all life on Earth.

Chapter 7: Challenges and Future Directions

A. Challenges Faced During the Expedition

1. Navigating Human-Influenced Landscapes

One of the most significant challenges faced during the expedition in the Alto Mayo region was the need to navigate landscapes heavily influenced by human activity. The encroachment of agriculture and deforestation has significantly altered the natural environment, reducing the availability of pristine habitats. According to Conservation International, approximately 25% of the forested area in the Alto Mayo landscape has been cleared for farming and urban expansion. These changes complicate efforts to locate and study species that rely on undisturbed ecosystems. For example, the swamp forest where the newly discovered amphibious mouse was found is under direct threat from agricultural practices, highlighting the tenuous balance between human development and biodiversity preservation.

Human settlements and agricultural zones also introduce invasive species that disrupt the balance of native ecosystems. Invasive plants and animals often compete with native species for resources, further endangering fragile populations. The team encountered areas where invasive grasses and crops had overrun native vegetation, making it harder to identify endemic plant species or the animals dependent on them.

Moreover, local infrastructure in these human-modified landscapes often lacks the resources to support scientific work. Roads are poorly maintained, and some regions are only accessible on foot or by canoe. This limits the researchers' ability to transport equipment and gather data efficiently. Despite these obstacles, the team managed to record over 2,000 species, a testament to their perseverance and strategic planning.

2. Limited Accessibility to Remote Habitats

Remote habitats in the Alto Mayo landscape presented another major challenge. While these areas often harbor the highest levels of biodiversity, they are typically difficult to access due to rugged terrain, dense forests, and minimal infrastructure. The region's varying topography—spanning the Andes mountains to the Amazon basin—necessitated extensive logistical planning and physical endurance from the research team.

For example, the amphibious mouse and blob-headed fish were found in isolated environments requiring specialized methods to reach. Swamp forests are particularly challenging due to their waterlogged soils and dense underbrush. Researchers had to navigate these areas on foot while carrying sensitive equipment, risking damage to tools essential for collecting samples and data.

Weather conditions further exacerbated accessibility issues. Rainfall in the Alto Mayo region can be unpredictable and intense, leading to flooding and erosion that hinder movement and increase the risk of accidents. The team had to time their surveys carefully, often working during short windows of favorable weather conditions.

Additionally, securing the safety of researchers in remote areas posed logistical difficulties. Wildlife encounters, such as with venomous snakes or large mammals, were a constant concern. Effective communication systems were essential but often unreliable due to the lack of cell towers and other infrastructure in remote regions. These challenges underscore the need for better support systems to facilitate future research.

B. Next Steps in Research and Conservation

1. Expanding Surveys to Other Regions

To build on the findings from the Alto Mayo landscape, expanding biodiversity surveys to other regions is crucial. Many areas in Peru and neighboring countries remain underexplored, particularly in transitional zones where ecosystems overlap. These regions are likely to harbor unique species adapted to niche environments, much like the amphibious mouse and blob-headed fish discovered during the expedition.

Future surveys should prioritize regions with high biodiversity potential but limited scientific documentation. For instance, cloud forests and other transitional ecosystems are known hotspots for species richness and endemism. Collaborative efforts between local universities, international organizations, and indigenous communities can help identify priority areas for exploration.

Innovative technologies such as remote sensing and environmental DNA (eDNA) analysis can also enhance survey efficiency. Remote sensing tools like satellite imagery and drones enable researchers to map difficult-to-reach areas and identify potential biodiversity hotspots. eDNA allows scientists to detect species presence through genetic material left in the environment, providing a non-invasive method to study elusive or rare organisms.

2. Securing Funding and Support for Long-Term Conservation

Long-term conservation efforts hinge on securing consistent funding and institutional support. The Alto Mayo landscape's biodiversity faces ongoing threats from deforestation, agricultural expansion, and climate change, necessitating sustained financial investment to address these challenges effectively.

One approach is to establish public-private partnerships that align conservation goals with economic incentives. For example, eco-tourism initiatives can generate revenue while promoting the preservation of natural habitats. Partnerships with local businesses can also support reforestation projects and sustainable agriculture programs.

International funding mechanisms such as the Global Environment Facility (GEF) and the Green Climate Fund (GCF) provide opportunities to scale up conservation efforts.

Additionally, crowdfunding platforms and social media campaigns can engage the global public, fostering a sense of shared responsibility for biodiversity conservation.

Education and capacity-building programs are equally important. Training local researchers and conservationists ensures that efforts are rooted in community knowledge and remain sustainable in the long term. For example, the involvement of the Awajún community in the Alto Mayo expedition highlights the value of indigenous knowledge in biodiversity research and conservation.

C. Raising Awareness

1. Communicating Findings to a Global Audience

Raising global awareness about the discoveries in the Alto Mayo region is essential for fostering broader support for conservation. Effective communication strategies should leverage traditional and digital media to reach diverse audiences.

Documentaries, articles, and social media campaigns can showcase the unique species discovered during the expedition, highlighting their ecological importance and the threats they face. Engaging visual content, such as photographs and videos of the blob-headed fish or amphibious mouse, can capture public interest and inspire action.

Collaborations with media outlets and influencers can amplify the message, ensuring that the expedition's findings reach international audiences. Educational institutions and museums can also play a role by incorporating these discoveries into exhibits and curricula, fostering interest in biodiversity among students and the general public.

2. Encouraging Policy Changes and Sustainable Practices

Advocating for policy changes at local, national, and international levels is critical for achieving lasting conservation outcomes. Policymakers must be informed about the ecological and economic benefits of preserving biodiversity, as well as the risks associated with inaction.

Legislation that supports sustainable land use and protects critical habitats is a key priority. For example, designating the swamp forest home of the amphibious mouse as a protected area could safeguard this unique ecosystem from agricultural encroachment. Similarly, policies that incentivize sustainable farming practices can reduce the impact of agriculture on biodiversity.

International agreements such as the Convention on Biological Diversity (CBD) provide frameworks for collaborative conservation efforts. Peru's participation in such agreements underscores its commitment to global biodiversity goals, but implementation requires continuous support and monitoring.

Community engagement is equally important for promoting sustainable practices. Programs that provide alternative livelihoods, such as agroforestry or eco-tourism, can help reduce dependence on activities that harm the environment. Indigenous communities, with their deep connection to nature, are vital allies in these efforts. The Awajún community's involvement in the Alto Mayo expedition exemplifies how indigenous knowledge and scientific research can work together to achieve conservation goals.

The challenges faced during the Alto Mayo expedition highlight the complexities of conducting biodiversity research in a rapidly changing world. However, these obstacles also underscore the urgency of expanding surveys, securing funding, and raising awareness to protect the planet's rich biological heritage. By addressing these challenges and embracing innovative approaches, we can pave the way for a future where biodiversity thrives alongside human development.

Conclusion

A. The Wonder of Discovery

The recent discovery of 27 new species in the Alto Mayo region of Peru stands as a testament to the vast, untapped mysteries of our planet. These findings—spanning mammals, fish, amphibians, and butterflies—underscore the incredible biodiversity that continues to thrive even in landscapes significantly influenced by human activity. This remarkable expedition not only enriches our scientific knowledge but also reignites our collective sense of wonder about the natural world.

Each of these discoveries tells a unique story about adaptation, survival, and the intricate interconnectedness of life. For instance, the amphibious mouse, with its webbed toes suited for aquatic habitats, exemplifies nature's ingenuity in responding to environmental challenges. Similarly, the "blob-headed" fish, with its peculiar and previously unseen head structure, raises questions about evolutionary adaptations that remain unexplained. These species, which have eluded scientific recognition until now, remind us of how little we truly know about the ecosystems around us.

The significance of these discoveries extends far beyond scientific curiosity. Each species plays a vital role in maintaining the balance of its ecosystem. Butterflies, for example, are essential pollinators that support plant reproduction and, by extension, the broader food web. Mammals such as the dwarf squirrel and short-tailed fruit bat contribute to seed dispersal and forest regeneration. Understanding these roles not only helps us appreciate the complexity of ecosystems but also highlights their vulnerability to disturbances.

Discovering new species also serves as a powerful reminder of the importance of preserving the unknown for future generations. These findings enrich our understanding of biodiversity and underscore the potential for unknown organisms to contribute to scientific advancements. From medicines derived from natural compounds to innovations inspired by biological processes, the benefits of biodiversity to humanity are profound.

Each newly discovered species represents an untapped reservoir of potential, emphasizing why protecting their habitats is not just an ethical obligation but also a pragmatic necessity.

Furthermore, these discoveries inspire a sense of awe and responsibility. They invite us to recognize the fragility and resilience of life in the face of mounting challenges such as deforestation, climate change, and habitat fragmentation. They compel us to act with urgency and foresight, ensuring that these species—and countless others yet to be discovered—can thrive in their natural environments.

In reflecting on these findings, it is crucial to acknowledge the collaborative efforts that made them possible. The involvement of indigenous communities, such as the Awajún, demonstrates the importance of integrating traditional ecological knowledge with modern scientific methods. This partnership not only enriches our understanding of biodiversity but also fosters a sense of shared stewardship over the natural world. By working together, we can ensure that the wonders of discovery continue to inspire and inform conservation efforts for generations to come.

B. A Call to Action

The discoveries in Peru's Alto Mayo region serve as a clarion call to individuals, communities, and governments worldwide. The responsibility of conserving biodiversity does not rest solely on scientists and environmental organizations; it is a collective endeavor that requires the participation of all stakeholders.

1. The Role of Individuals

Every individual has a role to play in conserving biodiversity. Simple actions, such as reducing waste, supporting sustainable products, and advocating for conservation policies, can collectively have a significant impact. Education and awareness are powerful tools for change. By learning about the importance of biodiversity and sharing this knowledge with others, individuals can foster a culture of conservation that permeates all levels of society.

Moreover, citizen science initiatives provide opportunities for people to contribute directly to conservation efforts. Whether it is participating in wildlife surveys, reporting sightings of rare species, or supporting local conservation projects, individuals can make meaningful contributions to preserving biodiversity. These actions not only help protect the environment but also deepen personal connections to the natural world, fostering a sense of stewardship that transcends generations.

2. The Role of Communities

Communities play a pivotal role in conservation, particularly those located near biodiversity hotspots. Local knowledge, such as that held by the Awajún people, is invaluable in understanding and protecting ecosystems. Indigenous communities, in particular, have long-standing relationships with their environments, possessing insights that are often overlooked by conventional scientific approaches.

Supporting community-led conservation initiatives is essential for achieving long-term sustainability. This includes providing resources for sustainable livelihoods, ensuring access to education and healthcare, and respecting indigenous rights and territories. When communities are empowered to manage their natural resources, they become effective custodians of biodiversity. Collaborative efforts between communities and conservation organizations can also bridge gaps in knowledge and foster innovative solutions to pressing environmental challenges.

3. The Role of Governments

Governments have a critical responsibility to implement policies and allocate resources for biodiversity conservation. This includes enforcing laws against illegal deforestation, poaching, and habitat destruction, as well as investing in protected areas and wildlife corridors. Governments must also prioritize sustainable development practices that balance economic growth with environmental preservation.

International cooperation is equally important. Biodiversity loss is a global issue that transcends national boundaries, requiring coordinated efforts to address its underlying causes. Treaties and agreements, such as the Convention on Biological Diversity, provide frameworks for countries to work together in conserving biodiversity and sharing its benefits equitably.

Furthermore, governments should support scientific research and innovation aimed at understanding and protecting biodiversity. Funding for expeditions, like the one in Alto Mayo, enables scientists to uncover new species and assess the health of ecosystems. These findings inform conservation strategies and contribute to achieving global biodiversity targets.

4. The Need for Continued Exploration

The discoveries in Alto Mayo underscore the importance of continued exploration and research. Despite significant advancements in technology and science, much of the natural world remains unexplored. Many species—particularly in remote and biodiverse regions—have yet to be documented, and their ecological roles remain unknown.

Investing in exploration is not merely an academic pursuit; it is a necessary step in addressing pressing environmental challenges. Understanding the distribution, behavior, and interactions of species is crucial for designing effective conservation measures. For example, identifying critical habitats and migration patterns allows for the creation of protected areas and wildlife corridors that support ecosystem connectivity.

The potential for discovery also extends to the field of bioprospecting, where researchers explore natural compounds for applications in medicine, agriculture, and industry. Many life-saving drugs, such as antibiotics and anticancer agents, have been derived from organisms found in biodiverse regions. Protecting these areas ensures that future discoveries can continue to benefit humanity.

5. The Urgency of Conservation

Time is of the essence in the fight to conserve biodiversity. The ongoing loss of species and habitats represents a crisis with far-reaching consequences for ecosystems and human well-being. Each extinction diminishes the resilience of ecosystems, making them more vulnerable to disturbances and less capable of providing essential services such as clean air, water, and food.

Achieving global biodiversity targets requires immediate and concerted action. This includes halting deforestation, restoring degraded ecosystems, and addressing the root causes of biodiversity loss, such as climate change and unsustainable resource use. Governments, organizations, and individuals must work together to implement solutions that balance environmental, social, and economic priorities.

The discoveries in Alto Mayo provide a source of hope and inspiration, demonstrating that it is not too late to make a difference. By acting now, we can safeguard the wonders of the natural world for future generations, ensuring that the planet's rich tapestry of life continues to thrive.

In conclusion, the discovery of 27 new species in Peru's Alto Mayo region is both a celebration of nature's wonders and a call to action for conservation. These findings highlight the importance of preserving the unknown, not only for its intrinsic value but also for the benefits it brings to humanity. By embracing a shared responsibility for the natural world, we can ensure that the beauty and complexity of life on Earth endure for generations to come. Let this moment of discovery inspire us to explore, protect, and cherish the biodiversity that sustains us all.

www.ingramcontent.com/pod-product-compliance
Lightning Source LLC
Chambersburg PA
CBHW050317230526
45471CB00005B/2219